The Incredible World of Insects

Incredible Ladybugs

By Susan Ashley

Gareth Stevens Publishing

Please visit our website, www.garethstevens.com. For a free color catalog of all our high-quality books, call toll free 1-800-542-2595 or fax 1-877-542-2596.

Library of Congress Cataloging-in-Publication Data

Ashley, Susan.
  Incredible ladybugs / Susan Ashley.
      p. cm. — (The incredible world of insects)
  Includes index.
  ISBN 978-1-4339-4592-2 (pbk.)
  ISBN 978-1-4339-4593-9 (6-pack)
  ISBN 978-1-4339-4591-5 (library binding)
  1. Ladybugs—Juvenile literature. I. Title.
  QL596.C65A743 2012
  595.76'9—dc22
                              2010035240

New edition published 2012 by
Gareth Stevens Publishing
111 East 14th Street, Suite 349
New York, NY 10003

New text and images this edition copyright © 2012 Gareth Stevens Publishing

Original edition published 2004 by Weekly Reader® Books
An imprint of Gareth Stevens Publishing
Original edition text and images copyright © 2004 Gareth Stevens Publishing

Designer: Daniel Hosek
Editors: Mary Ann Hoffman and Kristen Rajczak

Photo credits: Cover, pp. 1, 7, 15, 21 Shutterstock.com; pp. 5, 9, 11, 17 © Robert & Linda Mitchell; p. 13 Oxford Scientific/Photolibrary/Getty Images; p. 19 © Walt Anderson/Visuals Unlimited.

All rights reserved. No part of this book may be reproduced in any form without permission in writing from the publisher, except by a reviewer.

Printed in the United States of America

CPSIA compliance information: Batch #CR217260GS: For further information contact Gareth Stevens, New York, New York at 1-800-542-2595.

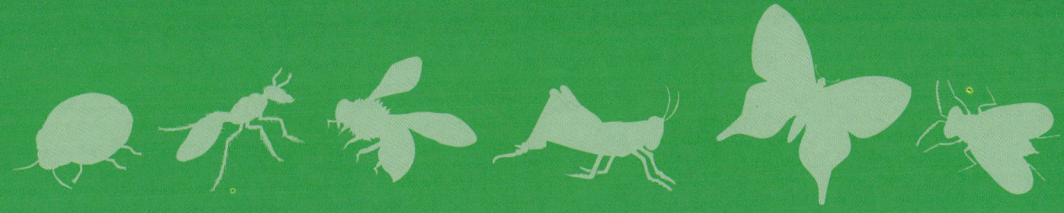

Contents

Look at Those Spots! 4
Beetles 6
From Egg to Ladybug 12
How Useful! 20
Glossary 22
For More Information 23
Index 24

Boldface words appear in the glossary.

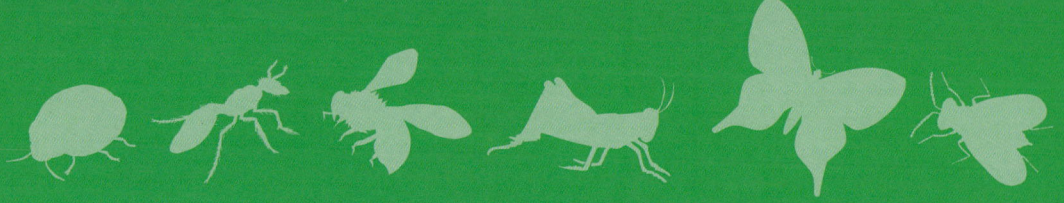

Look at Those Spots!

You've probably seen ladybugs. They are round **insects** with black spots. Some have two spots. Some have more. Many ladybugs are red. Others are orange or yellow.

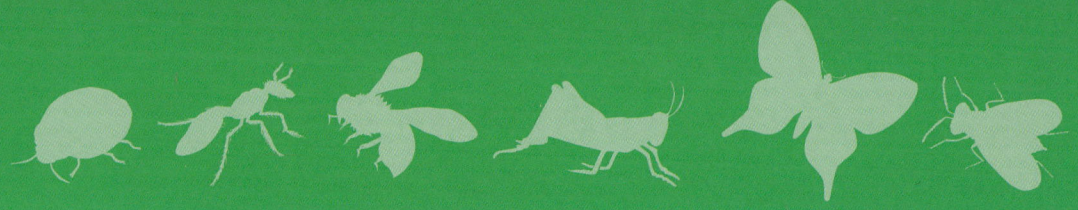

Beetles

Ladybugs are **beetles**. A ladybug has two hard front wings. These wings cover the ladybug's body. They also cover its two soft back wings.

7

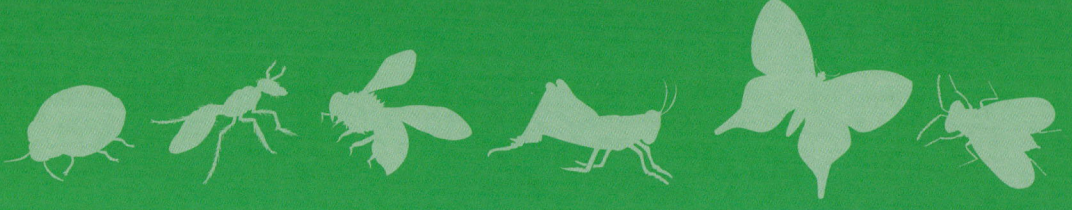

A ladybug's hard front wings spread apart when it flies. The soft back wings then move up and down.

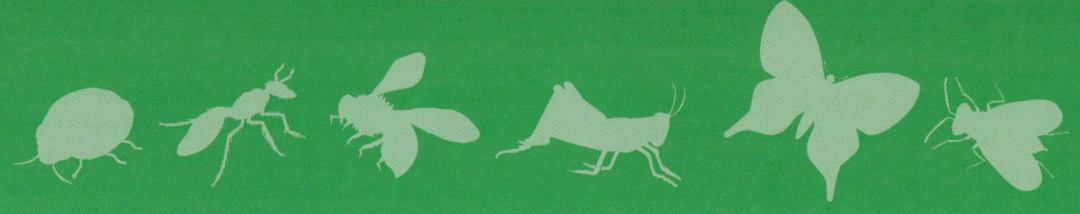

Ladybugs are not all ladies. There are female and male ladybugs. They look alike. Ladybugs **mate** in spring and summer.

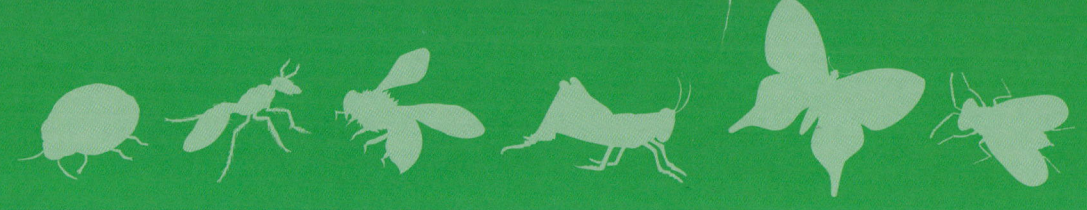

From Egg to Ladybug

The female ladybug lays eggs after mating. She lays many small yellow eggs on a leaf. The eggs **hatch**. A **larva** comes out of each egg.

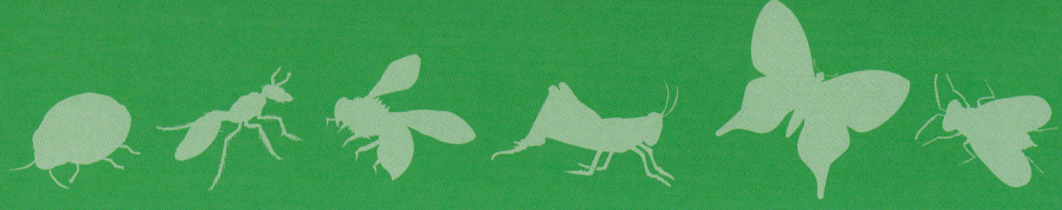

The larva has six legs. It looks much like a caterpillar. It eats and grows. It **sheds** its skin several times.

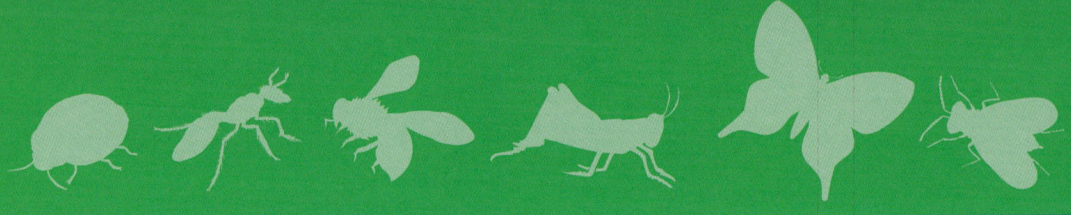

After a few weeks, the larva sticks itself to a leaf. It sheds its skin again and turns into a **pupa**. The pupa changes inside a hard case. It comes out a ladybug!

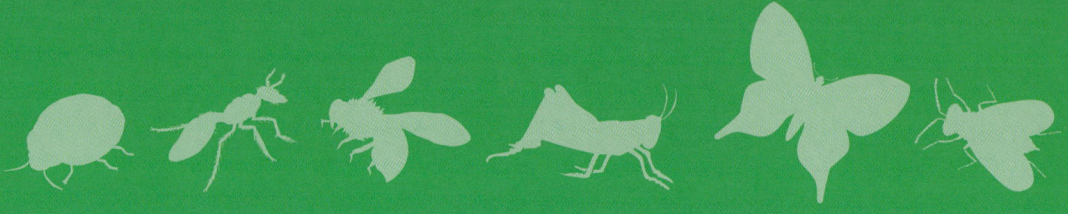

Ladybugs are very busy in summer. When it starts to get cold, they gather in groups. They find a safe place to **hibernate** until spring.

How Useful!

Ladybugs eat insects that harm trees and plants. That's why farmers and gardeners like ladybugs! Have you seen ladybugs in your yard? Count their spots!

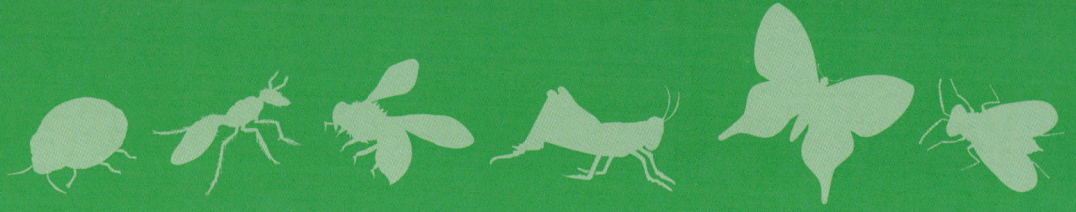

Glossary

beetle: an insect with two hard front wings and two soft back wings

hatch: to come out of an egg

hibernate: to pass the winter resting, without eating

insect: an animal with six legs, one or two pairs of wings, and three body parts

larva: the stage between egg and pupa

mate: to come together to make babies

pupa: the stage between larva and adult

shed: to get rid of

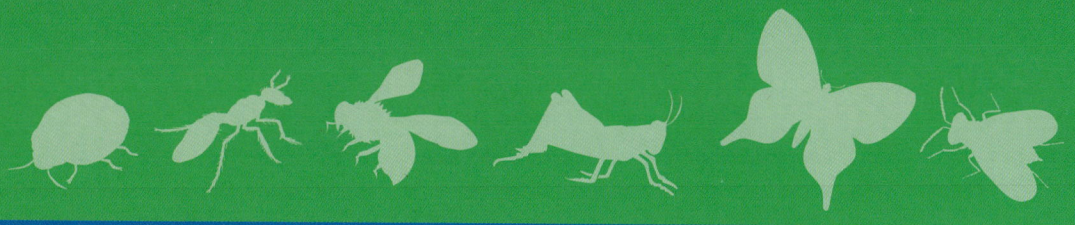

For More Information

Books

Rustad, Martha E. H. *Ladybugs*. Minneapolis, MN: Bellwether Media, 2008.

Smith, Molly. *Helpful Ladybugs*. New York, NY: Bearport Publishing, 2008.

Web Sites

Beetle
www.4to40.com/encyclopedia/index.asp?id=631
Read all about a group of insects called beetles.

Ladybug
animals.nationalgeographic.com/animals/bugs/ladybug.html
Find fast facts about ladybugs. View a world map of where they live.

Publisher's note to educators and parents: Our editors have carefully reviewed these websites to ensure that they are suitable for students. Many websites change frequently, however, and we cannot guarantee that a site's future contents will continue to meet our high standards of quality and educational value. Be advised that students should be closely supervised whenever they access the Internet.

Index

beetles 6
cold 18
eggs 12
farmers 20
female 10, 12
gardeners 20
hard case 16
hibernate 18
insects 4, 20
larva 12, 14, 16
male 10

mate 10, 12
orange 4
pupa 16
red 4
shed 14, 16
skin 14, 16
spots 4, 20
spring 10, 18
summer 10, 18
wings 6, 8
yellow 4